# The ESSENTIALS of
# THERMODYNAMICS II

**Staff of Research and Education Association,
Dr. M. Fogiel, Director**

> This book is a continuation of *"THE ESSENTIALS OF THERMODYNAMICS I"* and begins with Chapter 8. It covers the usual course outline of Thermodynamics II. Earlier/basic topics are covered in *"THE ESSENTIALS OF THERMODYNAMICS I"*.

Research and Education Association
505 Eighth Avenue
New York, N.Y. 10018

# THE ESSENTIALS OF THERMODYNAMICS II

Copyright © 1987 by Research and Education Association. All rights reserved. No part of this book may be reproduced in any form without permission of the publishers.

Printed in the United States of America

Library of Congress Catalog Card Number 87-61802

International Standard Book Number 0-87891-627-X

# WHAT "THE ESSENTIALS" WILL DO FOR YOU

This book is a review and study guide. It is comprehensive and it is concise.

It helps in preparing for exams, in doing homework, and remains a handy reference source at all times.

It condenses the vast amount of detail characteristic of the subject matter and summarizes the **essentials** of the field.

It will thus save hours of study and preparation time.

The book provides quick access to the important facts, principles, theorems, concepts, and equations of the field.

Materials needed for exams, can be reviewed in summary form — eliminating the need to read and re-read many pages of textbook and class notes. The summaries will even tend to bring detail to mind that had been previously read or noted.

This "ESSENTIALS" book has been carefully prepared by educators and professionals and was subsequently reviewed by another group of editors to assure accuracy and maximum usefulness.

Dr. Max Fogiel
Program Director

# CONTENTS

> This book is a continuation of "THE ESSENTIALS OF THERMODYNAMICS I" and begins with Chapter 8. It covers the usual course outline of Thermodynamics II. Earlier/basic topics are covered in "THE ESSENTIALS OF THERMODYNAMICS I".

**Chapter No.**             **Page No.**

## 8    THERMODYNAMIC RELATIONS    54

| | | |
|---|---|---|
| 8.1 | The Maxwell Relations | 54 |
| 8.2 | Clapeyron Equation | 54 |
| 8.3 | Other Thermodynamic Relations | 55 |
| 8.4 | Equations of State | 56 |
| 8.5 | Law of Corresponding States | 59 |
| 8.6 | Volume Expansion and Isothermal and Adiabatic Compressibility | 59 |

## 9    POWER AND REFRIGERATION CYCLES    61

| | | |
|---|---|---|
| 9.1 | Vapor Power Cycles | 61 |
| 9.1.1 | Simple Rankine Cycle | 61 |
| 9.1.2 | Rankine Cycle with Super Heater | 62 |
| 9.1.3 | The Reheat Cycle | 63 |

| | | |
|---|---|---|
| 9.1.4 | The Regenerative Cycle with Open Feedwater Heater | 64 |
| 9.2 | Deviation of Actual Cycles from Ideal Cycles | 65 |
| 9.2.1 | Piping Losses | 65 |
| 9.2.2 | Turbine Losses | 65 |
| 9.2.3 | Pump Losses | 66 |
| 9.2.4 | Condenser | 66 |
| 9.3 | Vapor Refrigeration Cycles | 66 |
| 9.4 | Air - Standard Power Cycles | 67 |
| 9.4.1 | The Carnot Cycle | 67 |
| 9.4.2 | The Otto Cycle | 68 |
| 9.4.3 | The Diesel Cycle | 69 |
| 9.4.4 | The Dual Cycle | 70 |
| 9.4.5 | Stirling Cycle | 71 |
| 9.4.6 | Ericsson Cycle | 71 |
| 9.4.7 | The Brayton Cycle | 72 |
| 9.4.8 | Brayton Cycle With Regenerator | 73 |
| 9.4.9 | Gas Refrigeration Cycle | 74 |

## 10  MIXTURES AND SOLUTIONS    75

| | | |
|---|---|---|
| 10.1 | Mole Fraction | 75 |
| 10.2 | Mass Fraction | 75 |
| 10.3 | Dalton's Rule of Partial Pressure | 75 |
| 10.4 | Amagat - Leduc Rule of Partial Volume | 76 |
| 10.5 | Expressions for Perfect Gases | 76 |
| 10.6 | Mixtures Involving Gases and Vapor | 77 |
| 10.6.1 | Dew Point Temperature | 78 |
| 10.7 | Enthalpy and Entropy of a Gas - Vapor Mixture | 79 |
| 10.8 | Adiabatic Saturation Process | 79 |
| 10.9 | Fugacity | 80 |
| 10.10 | Activity | 81 |
| 10.10.1 | Activity Coefficient | 81 |

## 11  CHEMICAL REACTIONS    82

| | | |
|---|---|---|
| 11.1 | The Combustion Process | 82 |
| 11.2 | Enthalpy of Reaction | 83 |

| | | |
|---|---|---|
| 11.3 | Enthalpy of Formation ($h_f^\circ$) | 83 |
| 11.4 | First Law Analysis of Reacting Systems | 84 |
| 11.5 | Adiabatic Flame Temperature | 85 |
| 11.6 | The Third Law of Thermodynamics and Absolute Entropy | 85 |
| 11.7 | Second Law Analysis of Reacting Systems | 86 |
| 11.8 | Gibb's Function | 87 |
| 11.9 | Chemical Potential | 87 |

| | | |
|---|---|---|
| **12** | **CHEMICAL EQUILIBRIUM** | **89** |
| 12.1 | Requirements for Chemical Equilibrium | 89 |
| 12.2 | Equilibrium and Chemical Potential | 89 |
| 12.3 | Equilibrium Constant of a Reactive Mixture of Ideal Gases | 90 |
| 12.4 | Equilibrium Between Two Phases of a Pure Substance | 91 |
| 12.5 | Equilibrium of a Multicomponent, Multiphase System | 91 |
| 12.6 | Gibb's Phase Rule | 92 |

| | | |
|---|---|---|
| **13** | **FLOW THROUGH NOZZLES AND BLADE PASSAGES** | **93** |
| 13.1 | Conservative of Mass for the Control Volume | 93 |
| 13.1.1 | Special Cases | 93 |
| 13.2 | Momentum Equation for the Control Volume | 93 |
| 13.3 | Speed of Sound | 94 |
| 13.3.1 | Mach Number | 94 |
| 13.4 | Local Isentropic Stagnation Properties | 94 |
| 13.5 | Critical Constants | 95 |
| 13.6 | Effects of Area Variation on Flow Properties In Isentropic Flow | 95 |
| 13.7 | Isentropic Flow of an Ideal Gas | 96 |

| | | |
|---|---|---|
| 13.8 | Isentropic Flow in a Converging and a Converging/Diverging Nozzle | 97 |
| 13.8.1 | Converging Nozzle | 97 |
| 13.8.2 | Converging/Diverging Nozzle | 98 |
| 13.9 | Normal Shocks | 98 |
| 13.9.1 | Flow in a Converging/Diverging Nozzle | 99 |
| 13.10 | Nozzle and Diffuser Coefficients | 100 |
| 13.11 | Flow Through Blade Passages | 101 |
| 13.12 | Impulse and Reaction Stages for Turbines | 103 |
| 13.12.1 | Impulse Stage | 103 |
| 13.12.2 | Reaction Stage | 104 |

# CHAPTER 8

# THERMODYNAMIC RELATIONS

## 8.1 THE MAXWELL RELATIONS

The Maxwell Relations are:

$$\left(\frac{\partial S}{\partial V}\right)_T = \left(\frac{\partial p}{\partial T}\right)_V \tag{8.1}$$

$$\left(\frac{\partial T}{\partial V}\right)_S = -\left(\frac{\partial p}{\partial S}\right)_V \tag{8.2}$$

$$\left(\frac{\partial S}{\partial p}\right)_T = -\left(\frac{\partial V}{\partial T}\right)_p \tag{8.3}$$

$$\left(\frac{\partial T}{\partial p}\right)_S = \left(\frac{\partial V}{\partial S}\right)_p \tag{8.4}$$

These relations are important because they relate the entropy to easily measured properties - pressure, volume and temperature.

## 8.2 CLAPEYRON EQUATION

The Clapeyron Equation is an important relation involving the saturation pressure and temperature, the change of enthalpy associated with a change of phase, and the specific volumes of the two phases.

The form of the Clapeyron equation is:

$$\frac{dp}{dT} = \frac{h_{fg}}{Tv_{fg}} \qquad (8.5)$$

where $h_{fg}$ is the enthalpy of vaporization.

## 8.3 OTHER THERMODYNAMIC RELATIONS

a) There are a number of other useful relations that can be easily derived:

$$\left(\frac{\partial U}{\partial S}\right)_V = T \qquad (8.6)$$

$$\left(\frac{\partial U}{\partial V}\right)_S = -p \qquad (8.7)$$

$$\left(\frac{\partial H}{\partial S}\right)_P = T \qquad (8.8)$$

$$\left(\frac{\partial H}{\partial p}\right)_S = V \qquad (8.9)$$

$$\left(\frac{\partial \psi}{\partial T}\right)_V = -S \qquad (8.10)$$

$$\left(\frac{\partial \psi}{\partial V}\right)_T = -p \qquad (8.11)$$

$$\left(\frac{\partial Z}{\partial T}\right)_P = -S \qquad (8.12)$$

$$\left(\frac{\partial Z}{\partial p}\right)_T = V \qquad (8.13)$$

b) Enthalpy

$$dh = C_p dT + \left[v - T\left(\frac{\partial v}{\partial T}\right)_p\right] dp \quad (8.14)$$

c) Internal Energy

$$du = C_v dT + \left[T\left(\frac{\partial p}{\partial T}\right)_v - p\right] dv \quad (8.15)$$

d) Entropy

$$ds = C_v \frac{dT}{T} + \left(\frac{\partial p}{\partial v}\right)_T dv \quad (8.16)$$

e) Specific heats

$$C_v = \left(\frac{\partial u}{\partial T}\right)_v = T\left(\frac{\partial S}{\partial T}\right)_v \quad (8.17)$$

$$C_p = \left(\frac{\partial h}{\partial T}\right)_p = T\left(\frac{\partial S}{\partial T}\right)_p \quad (8.18)$$

$$\left(\frac{\partial C_v}{\partial v}\right)_T = T\left(\frac{\partial^2 p}{\partial T^2}\right)_v \quad (8.19)$$

$$\left(\frac{\partial C_p}{\partial p}\right)_T = -T\left(\frac{\partial^2 v}{\partial T^2}\right)_p \quad (8.20)$$

## 8.4 EQUATIONS OF STATE

The P-V-T relation is often stated in the form of an equation which is called an equation of state:

$$PV = RT \quad (8.21)$$

a) Van der Waals Equation

$$p = \frac{RT}{v-b} - \frac{a}{v^2} \quad (8.22)$$

a, b are constants for any one substance.

b) Dieterici Equation

$$p = \frac{RT}{v-b} \exp\left(-\frac{a}{vRT}\right) \qquad (8.23)$$

c) Beattie-Bridgeman Equation

$$p = \frac{RT(1-\varepsilon)}{v^2}(v+B) - \frac{A}{v^2} \qquad (8.24)$$

$A = A_0(1-a/v)$, $B = B_0(1-B/v)$, $\varepsilon = c/UT^3$ and $A_0$, $a$, $B_0$, $b$ and $c$ are constants for different gases. Values for constants are given in table 1.

d) Benedict-Webb-Rubin Equation

$$p = \frac{RT}{v} + \left(B_0 RT - A_0 - \frac{C_0}{T^2}\right)\frac{1}{v^2} + (bRT-a)\frac{1}{v^3} + \frac{a\alpha}{v^6}$$

$$+ \frac{c\left(1+\frac{\gamma}{v^2}\right)}{T^2} \cdot \frac{1}{v^3} \exp\left(-\frac{\gamma}{v^2}\right) \qquad (8.25)$$

Values for the constants are given in table 2.

Constants of the Beattie-Bridgeman Equation of State

$$R = 0.08206 \frac{(atm)(liters)}{(g\text{-mole})(°K)}$$

| Gas | $A_0$ | a | $B_0$ | b | $10^{-4}c$ |
|---|---|---|---|---|---|
| Helium | 0.0216 | 0.05984 | 0.01400 | 0.0 | 0.0040 |
| Neon | 0.2125 | 0.02196 | 0.02060 | 0.0 | 0.101 |
| Argon | 1.2907 | 0.02328 | 0.03931 | 0.0 | 5.99 |
| Hydrogen | 0.1975 | -0.00506 | 0.02096 | -0.04359 | 0.0504 |
| Nitrogen | 1.3445 | 0.02617 | 0.05046 | -0.00691 | 4.20 |
| Oxygen | 1.4911 | 0.02562 | 0.04624 | 0.004208 | 4.80 |
| Air | 1.3012 | 0.01931 | 0.04611 | -0.001101 | 4.34 |
| $CO_2$ | 5.0065 | 0.07132 | 0.10476 | 0.07235 | 66.00 |
| $(C_2H_5)_2O$ | 31.278 | 0.12426 | 0.45446 | 0.11954 | 33.33 |
| $C_2H_4$ | 6.152 | 0.04964 | 0.12156 | 0.03597 | 22.68 |
| Ammonia | 2.3930 | 0.17031 | 0.03415 | 0.19112 | 476.87 |
| CO | 1.3445 | 0.02617 | 0.05046 | -0.00691 | 4.20 |
| $N_2O$ | 5.0065 | 0.07132 | 0.10476 | 0.07235 | 66.0 |
| $CH_4$ | 2.2769 | 0.01855 | 0.05587 | -0.01587 | 12.83 |
| $C_2H_6$ | 5.8800 | 0.05861 | 0.09400 | 0.01915 | 90.00 |
| $C_3H_8$ | 11.9200 | 0.07321 | 0.18100 | 0.04293 | 120 |
| $n-C_4H_{10}$ | 17.794 | 0.12161 | 0.24620 | 0.09423 | 350 |
| $n-C_7H_{16}$ | 54.520 | 0.20066 | 0.70816 | 0.19179 | 400 |

Table 1

Empirical Constants for Benedict-Webb-Rubin Equation

Units: Atmospheres, liters, moles, °K.  Gas Constants: $R = 0.08207$; $T = 273.13 + t(°C)$

| Gas | $A_0$ | $B_0$ | $C_0 \times 10^{-6}$ | a | b | $c \times 10^{-6}$ | $\alpha \times 10^3$ | $\gamma \times 10^2$ |
|---|---|---|---|---|---|---|---|---|
| Nitrogen | 1.19250 | 0.0458000 | 0.00588907 | 0.0149000 | 0.00198154 | 0.000548064 | 0.291545 | 0.750000 |
| Methane | 1.85500 | 0.0426000 | 0.0225700 | 0.494000 | 0.00338004 | 0.00254500 | 0.124359 | 0.60000 |
| Ethylene | 3.33958 | 0.0556833 | 0.131140 | 0.259000 | 0.0086000 | 0.021120 | 0.178000 | 0.923000 |
| Ethane | 4.15556 | 0.0627724 | 0.179592 | 0.345160 | 0.0111220 | 0.0327670 | 0.243389 | 1.18000 |
| Propylene | 6.11220 | 0.0850647 | 0.439182 | 0.774056 | 0.0187059 | 0.102611 | 0.455696 | 1.82900 |
| Propane | 6.87225 | 0.0973130 | 0.508256 | 0.947700 | 0.0225000 | 0.129000 | 0.607175 | 2.20000 |
| i-Butane | 10.23264 | 0.137544 | 0.849943 | 1.93763 | 0.0424352 | 0.286010 | 1.07408 | 3.40000 |
| i-Butylene | 8.95325 | 0.116025 | 0.927280 | 1.69270 | 0.0348156 | 0.274920 | 0.910889 | 2.95945 |
| n-Butane | 10.0847 | 0.124361 | 0.992830 | 1.88231 | 0.0399983 | 0.316400 | 1.10132 | 3.40000 |
| i-Pentane | 12.7959 | 0.160053 | 1.74632 | 3.75620 | 0.0668120 | 0.695000 | 1.70000 | 4.63000 |
| n-Pentane | 12.1794 | 0.156751 | 2.12121 | 4.07480 | 0.0668120 | 0.824170 | 1.81000 | 4.75000 |
| n-Hexane | 14.4373 | 0.177813 | 3.31935 | 7.11671 | 0.109131 | 1.51276 | 2.81086 | 6.66849 |
| n-Heptane | 17.5206 | 0.199005 | 4.74574 | 10.36475 | 0.151954 | 2.47000 | 4.35611 | 9.00000 |

Table 2

e) Serial Forms of the Equations of State

The P-V-T relation may be expressed in terms of an infinite series in powers of the density as follows:

$$z = \frac{Pv}{RT} = 1 + \frac{B(T)}{v} + \frac{C(T)}{v^2} + \frac{D(T)}{v^3} + \ldots \qquad (8.26)$$

where $B(T)$, $C(T)$, $D(T)$ are called virial coefficients.

Any of the equations of state discussed above may be expanded into a series form similar to (8.26). For example, the Van der Waals equation yields simply:

$$B(T) = b - \frac{a}{RT} \qquad (8.27)$$

$$B(T) = b^2$$

$$C(T) = b^3 \quad \text{etc.}$$

## 8.5 LAW OF CORRESPONDING STATES

The law of corresponding states declares that there is a single functional relationship of the form,

$$v_R = f(p_R, T_R),$$

which holds for all substances.

## 8.6 VOLUME EXPANSION AND ISOTHERMAL AND ADIABATIC COMPRESSIBILITY

a) Coefficient of linear expansion

$$\delta_P = \frac{1}{L}\left(\frac{\partial L}{\partial T}\right)_P \qquad (8.28)$$

b) Volume

$$\alpha_P = \frac{1}{V}\left(\frac{\partial V}{\partial T}\right)_P = \frac{1}{v}\left(\frac{\partial V}{\partial T}\right)_P \tag{8.29}$$

c) Isothermal compressibility

$$\boxed{\beta_T = -\frac{1}{V}\left(\frac{\partial V}{\partial P}\right)_T = -\frac{1}{v}\left(\frac{\partial v}{\partial P}\right)_T} \tag{8.30}$$

d) Adiabatic compressibility

$$\boxed{\beta_S = -\frac{1}{v}\left(\frac{\partial v}{\partial P}\right)_S} \tag{8.31}$$

e) Adiabatic bulk modulus

$$\boxed{\beta_S = -v\left(\frac{\partial P}{\partial v}\right)_S} \tag{8.32}$$

# CHAPTER 9

# POWER AND REFRIGERATION CYCLES

## 9.1 VAPOR POWER CYCLES

### 9.1.1 SIMPLE RANKINE CYCLE

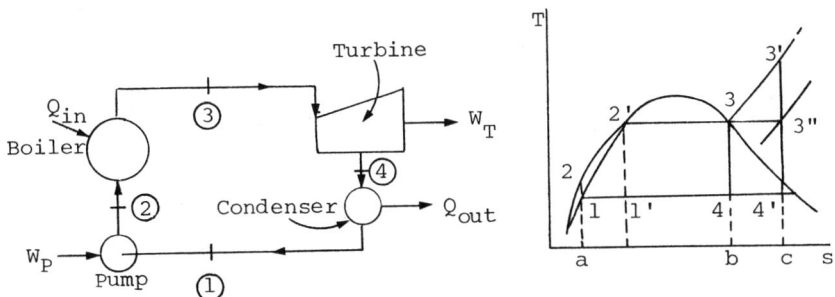

Fig. 9.1 Simple steam power plant which operates on the Rankine cycle.

A Rankine cycle consists of the following processes:

1-2 Reversible adiabatic pumping process in the pump

2-3 Constant-pressure transfer of heat in the boiler

3-4 Reversible adiabatic expansion in the turbine

4-1 Constant pressure transfer of heat in the condenser

Assuming steady-state, steady-flow processes throughout and neglecting changes in kinetic energy and potential energy across each piece of cycle component, we have, from the first law, the following results:

Boiler: $Q_{in} = h_3 - h_2$

Turbine: $W_T = h_3 - h_4$

Condenser: $Q_{out} = h_4 - h_1$

Pump: $W_p = h_2 - h_1 = v(P_2 - P_1)$

The thermal efficiency can be expressed in terms of properties at various points in the cycle:

$$n_{th} = \frac{W_{net}}{Q_{in}} = \frac{W_T - W_p}{Q_{in}} = \frac{(h_3 - h_2) - (h_4 - h_1)}{h_3 - h_2} = \frac{(h_3 - h_4) - (h_2 - h_1)}{h_3 - h_2}$$

(9.1)

The efficiency of the cycle can be increased by lowering the exhaust pressure, increasing the pressure during heat addition, or superheating the steam.

## 9.1.2 RANKINE CYCLE WITH SUPER HEATER

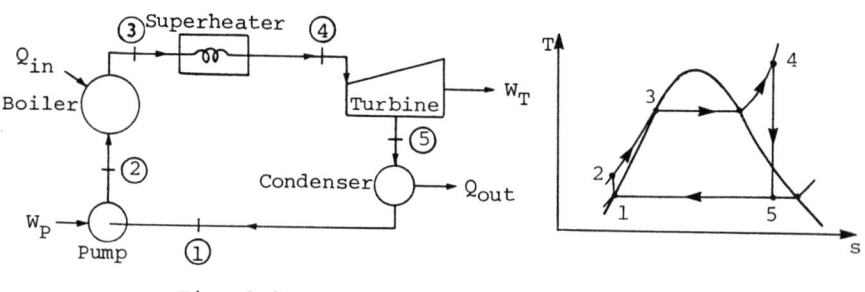

Fig. 9.2

We may operate a Rankine cycle with superheat as shown in Fig. 9.2. Using this method, we achieve a higher mean temperature of heat addition without increasing the maximum pressure of the cycle.

For this cycle we have:

$$Q_{in} = h_4 - h_2$$

$$W_T = h_4 - h_5$$

$$Q_{out} = h_5 - h_1$$

$$W_P = h_2 - h_1 = v_1(P_2 - P_1) \quad \text{and}$$

$$\boxed{n_{th} = \frac{(h_4 - h_5) - (h_2 - h_1)}{h_4 - h_2}} \quad (9.2)$$

## 9.1.3 THE REHEAT CYCLE

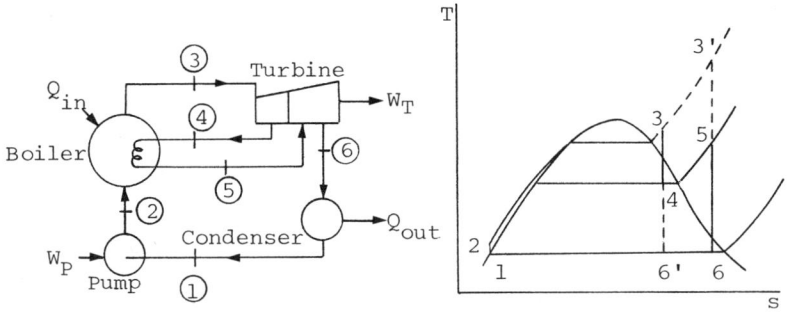

Fig. 9.3 The ideal reheat cycle

The reheat cycle has been developed to take advantage of the increased efficiency associated with higher pressures.

The turbine may be considered as having two stages: high-pressure and low-pressure. For this cycle we have:

$$Q_{in} = (h_3 - h_2) + (h_5 - h_4)$$

$$W_T = (h_3 - h_4) + (h_5 - h_6)$$

$$Q_{out} = (h_6-h_1)$$

$$W_P = (h_2-h_1) = v_1(P_2-P_1)$$

$$\boxed{n_{th} = \frac{(h_3-h_4)+(h_5-h_6)-(h_2-h_1)}{(h_3-h_2)+(h_5-h_4)}} \quad (9.3)$$

## 9.1.4 THE REGENERATIVE CYCLE WITH OPEN FEEDWATER HEATER

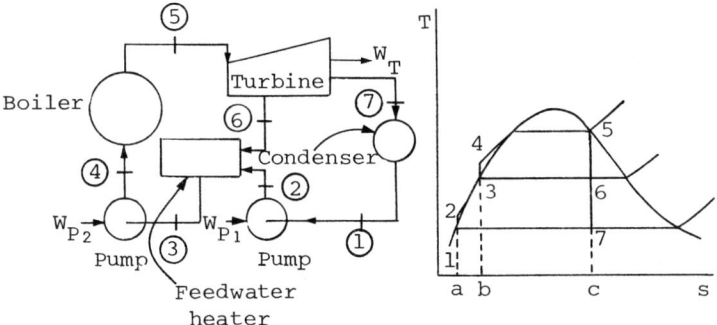

Fig. 9.4 Regenerative cycle with open feedwater heater.

This regenerative cycle involves the extraction of some of the vapor after it has partially expanded, and the use of feedwater heaters (Fig. 9.4). The number of stages of extraction is determined by economic considerations. For this cycle we have:

$$Q_{in} = h_5 - h_4$$

$$W_T = (h_5-h_6)+(1-m_1)(h_6-h_7)$$

$$Q_{out} = (h_1-h_7)$$

$$W_{P_2} = (h_4-h_3) = v(P_4-P_3)$$

$$W_{P_1} = (h_2-h_1) = v(P_2-P_1)$$

$$n_{th} = \frac{W_T - (1-m_1)W_{P_1} - W_{P_2}}{(h_5-h_4)} \quad (9.4)$$

Around the feedwater heater we have:

$$m_1(h_6) + (1-m_1)h_2 = h_3$$

## 9.2 DEVIATION OF ACTUAL CYCLES FROM IDEAL CYCLES

The most important reasons for the deviation of actual cycles from ideal cycles are:

### 9.2.1 PIPING LOSSES

Observed pressure drops are due to frictional effects and heat transfer to the surroundings. The heat transfer causes a decrease in entropy. Also, both the pressure drops and heat transfer cause a decrease in the availability.

### 9.2.2 TURBINE LOSSES

The losses in the turbine are associated with the flow of the working fluid. The effects are the same as those outlined for piping losses.

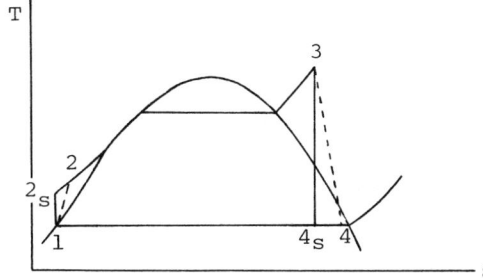

Fig. 9.5  Temperature-entropy diagram showing effect of turbine and pump inefficiencies on cycle performance.

The efficiency of the turbine has been defined,

$$\boxed{n_t = \frac{W_t}{h_3 - h_{4s}}} \quad (9.5)$$

4 represents the actual state leaving the turbine and 4s represents the state after an isentropic expansion.

### 9.2.3 PUMP LOSSES

The losses in the pump are similar to those of the turbine. The pump efficiency is defined as

$$n_p = \frac{h_{2s} - h}{W_p} \quad (9.6)$$

### 9.2.4 CONDENSER

The losses in the condenser are relatively small. One of these is the cooling below saturation temperature of the liquid leaving the condenser.

## 9.3 VAPOR REFRIGERATION CYCLES

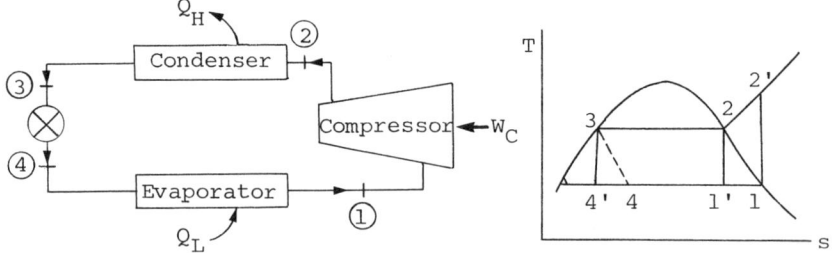

Fig. 9.6 The ideal vapor-compression refrigeration cycle.

The vapor refrigeration cycle is essentially the same as the Rankine cycle in reverse. The only difference is that the expansion valve replaces the pump. The performance of a regrigerator cycle is given as:

$$\boxed{\beta = \frac{Q_L}{W_c} = \frac{h_1 - h_4}{h_2 - h_1}} \qquad (9.7)$$

$$\text{capacity (tons)} = \frac{h_1 - h_4}{12,000 \text{ Btu/hr}} \qquad (9.8)$$

## 9.4 AIR-STANDARD POWER CYCLES

An Air-standard cycle operates under the following assumptions:

1. A fixed mass of air is the working fluid and the air is always an ideal gas.
2. The combustion process is replaced by a heat transfer process from an external source.
3. The cycle is completed by heat transfer to the surroundings.
4. All processes are internally reversible.
5. Air has a constant specific heat.

### 9.4.1 THE CARNOT CYCLE

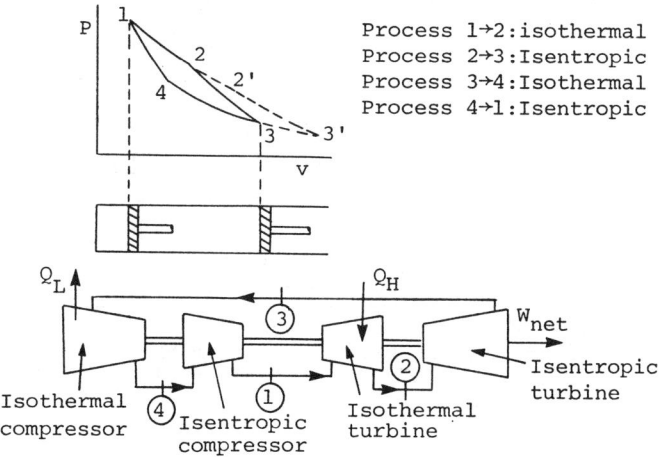

Process 1→2: isothermal
Process 2→3: Isentropic
Process 3→4: Isothermal
Process 4→1: Isentropic

Fig. 9.7 The air-standard Carnot cycle.

The thermal efficiency of the Carnot cycle is:

$$n_{th} = 1 - \frac{T_L}{T_H} = 1 - \frac{T_4}{T_1} = 1 - \frac{T_3}{T_2} \quad (9.9)$$

The efficiency may also be expressed by the pressure ratio or compression ratio.

Isentropic pressure ratio: $r_{ps} = \frac{P_1}{P_4} = \frac{P_2}{P_3} = \left(\frac{T_3}{T_2}\right)^{K/(1-K)}$ (9.10)

Isentropic compression ratio:

$$r_{vs} = \frac{V_4}{V_1} = \frac{V_3}{V_2} = \left(\frac{T_3}{T_2}\right)^{1/(1-K)} \quad (9.11)$$

$$\therefore n_{th} = 1 - r_{ps}^{(1-K)/K} = 1 - r_{vs}^{1-K} \quad (9.12)$$

## 9.4.2 THE OTTO CYCLE

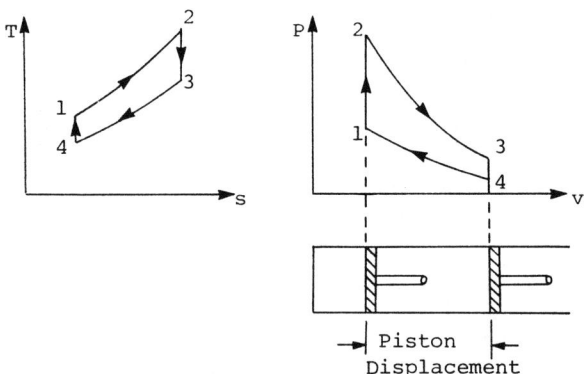

1. Constant-volume heat addition (process 1→2).
2. Isentropic expansion (process 2→3).
3. Constant-volume heat rejection (process 3→4).
4. Isentropic compression (process 4→1).

Fig. 9.8 Otto Cycle

On the basis of unit mass of gas, we have for the Otto cycle:

$$Q_{in} = Q_{12} = C_v(T_2-T_1)$$

$$Q_{out} = U_{34} = C_v(T_3-T_4)$$

$$n_{th} = \frac{W_{net}}{Q_{in}} = \frac{Q_{in}-Q_{out}}{Q_{in}}$$

$$\therefore \quad n_{th} = 1 - \frac{(T_3-T_4)}{(T_2-T_1)} = 1 - \frac{1}{r_v^{(K-1)}} \quad (9.13)$$

where $r_v = v_3/v_2 = v_4/v_1$ is known as the compression rate of the cycle

Note: The Otto cycle efficiency increases with increased compression ratio.

## 9.4.3 THE DIESEL CYCLE

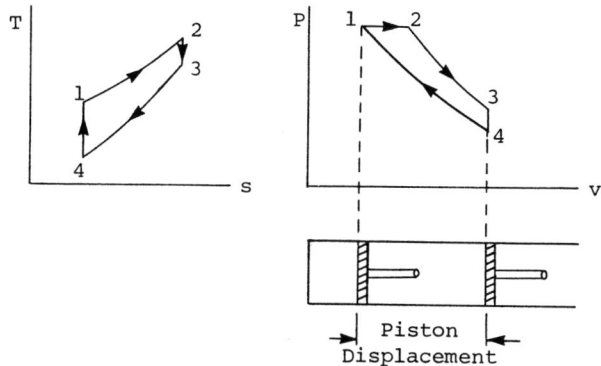

1. Constant-pressure heat addition (process 1→2).
2. Isentropic expansion (process 2→3).
3. Constant-volume heat rejection (process 3→4).
4. Isentropic compression (process 4→1).

Fig. 9.9 Diesel Cycle

On the basis of unit mass of gas, we have for the Diesel cycle:

$$Q_{in} = Q_{12} = C_p(T_2-T_1)$$

$$Q_{out} = Q_{34} = C_v(T_3-T_4)$$

$$n_{th} = \frac{Q_{in} - Q_{out}}{Q_{in}} = 1 - \frac{(T_3-T_4)}{K(T_2-T_1)}$$

∴
$$n_{th} = 1 - \frac{1}{r_v^{K-1}} \cdot \frac{r_c^K - 1}{K(r_c-1)} \qquad (9.14)$$

$z_v = \frac{V_4}{V_1}$ is the compression ratio.

$z_c = \frac{V_2}{V_1}$ is the cutoff ratio.

Note: The efficiency of a Diesel cycle is always lower than that of an Otto cycle with the same compression ratio.

## 9.4.4 THE DUAL CYCLE

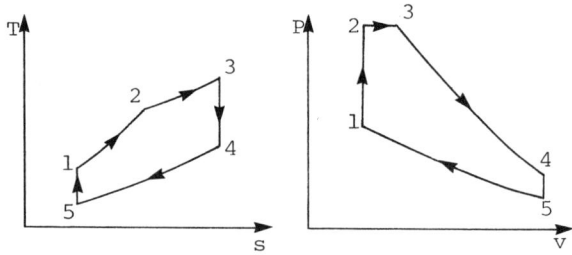

Fig. 9.10 Dual Cycle

The Dual cycle is a result of combining the processes of the Otto and Diesel cycles.

On the basis of unit mass, we have for the dual cycle:

$$Q_{in} = Q_{12} + Q_{23} = Cv(T_2-T_1) + Cp(T_3-T_2)$$

$$Q_{out} = Q_{45} = Cv(T_5-T_4)$$

$$n_{th} = 1 - \frac{T_4 - T_5}{(T_2-T_1) + K(T_3-T_2)}$$

∴
$$n_{th} = 1 - \frac{1}{r_v^{K-1}} \left[ \frac{r_p r_c^K - 1}{Kr_p(r_c-1) + r_p - 1} \right] \qquad (9.15)$$

When $r_p = 1$, equation 9.15 reduces to the Diesel efficiency equation:

$$r_v = \frac{V_5}{V_1} \quad , \quad r_c = \frac{V_3}{V_2} \quad , \quad r_p = \frac{P_2}{P_1}$$

## 9.4.5 STIRLING CYCLE

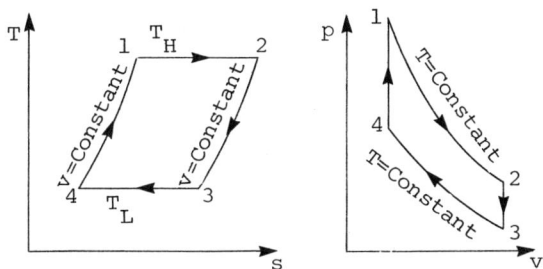

Fig. 9.11 Stirling Cycle

For a Stirling cycle, we have, on the basis of unit mass:

$$Q_{in} = Q_{12} = T_H(s_2 - s_1) = RT_H \ln \frac{V_2}{V_1} = -RT_H \ln \frac{P_2}{P_1}$$

$$Q_{out} = Q_{34} = T_L(s_3 - s_4) = RT_L \ln \frac{V_4}{V_3} = -RT_L \ln \frac{P_2}{P_1}$$

$$\boxed{n_{th} = \frac{T_H - T_L}{T_H}} \qquad (9.16)$$

## 9.4.6 ERICSSON CYCLE

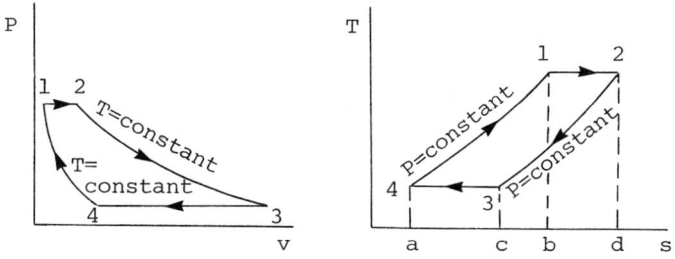

Fig. 9.12 The air-standard Ericsson cycle.

The constant-volume processes of the Stirling cycle are replaced by constant-pressure processes in the Ericsson cycle.

## 9.4.7 THE BRAYTON CYCLE

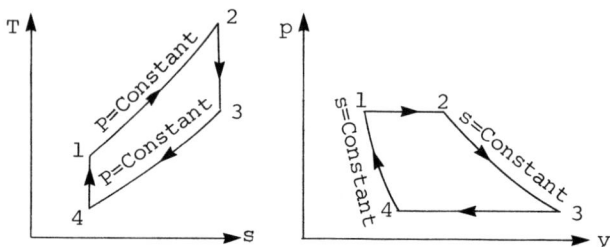

Fig. 9.13  Brayton Cycle

For the air-standard Brayton cycle, we have on the basis of unit mass:

$$Q_{in} = Q_{12} = C_p(T_2-T_1)$$

$$Q_{out} = Q_{34} = C_p(T_4-T_3)$$

$$\boxed{n_{th} = 1 - \frac{T_3-T_4}{T_2-T_1} = 1 - \frac{1}{r_p^{(K-1)/K}}} \qquad (9.17)$$

$r_p = p_1/p_4 = p_2/p_3$ is known as the pressure ratio.

Two general types of gas-turbine power plants have been developed based on the Brayton cycle: closed-cycle and open-cycle.

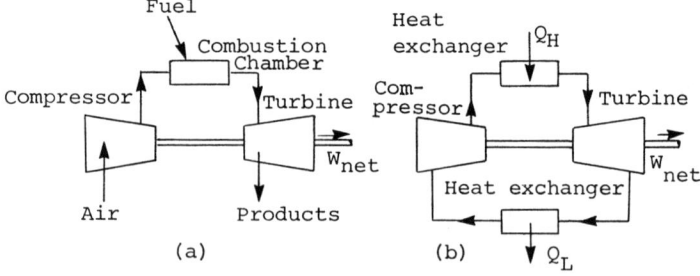

Fig. 9.14  A gas turbine operating on the Brayton cycle. (a) Open cycle. (b) Closed cycle.

## 9.4.8 BRAYTON CYCLE WITH REGENERATOR

Brayton Cycle with Regenerator

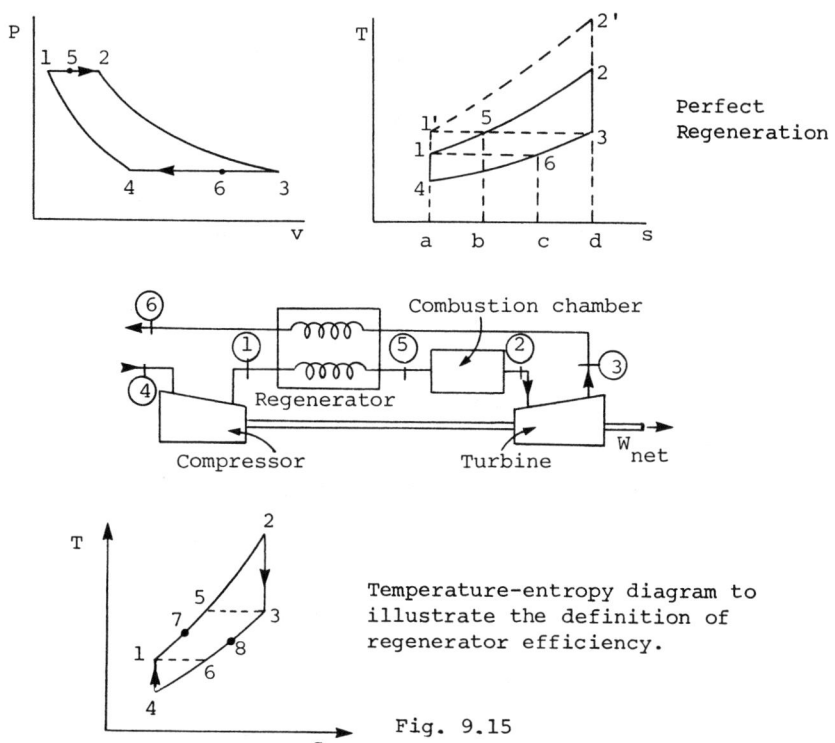

Perfect Regeneration

Temperature-entropy diagram to illustrate the definition of regenerator efficiency.

Fig. 9.15

For this cycle we have:

$$W_{net} = C_p(T_2-T_3) - C_p(T_2-T_4)$$

$$Q_{in} = Q_{52} = C_p(T_2-T_5) = C_p(T_2-T_3)$$

$$n_{th} = \frac{(T_2-T_3)-(T_1-T_4)}{(T_2-T_3)} = 1 - \frac{T_4}{T_2}\left(\frac{p_1}{p_4}\right)^{(K-1)/K} \qquad (9.18)$$

In this case, the heat absorbed by the air leaving the compressor is identical to the heat given up by the gas leaving the turbine.

The regenerator effectiveness is defined:

$$n_{reg} = \frac{T_7 - T_1}{T_3 - T_1} \qquad (9.19)$$

## 9.4.9 GAS REFRIGERATION CYCLE

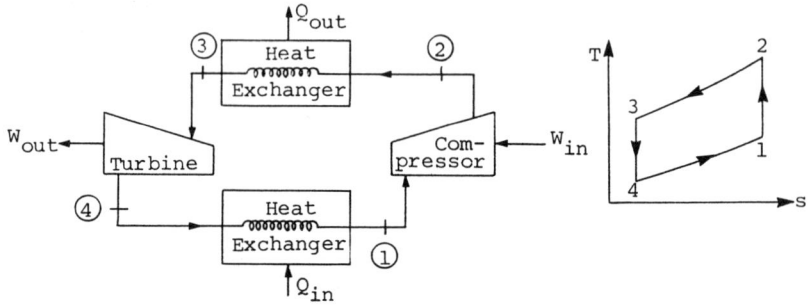

Fig. 9.16 Gas Refrigeration Cycle

1. Isentropic compression (process 1-2)
2. Cooling at constant pressure (process 2-3)
3. Isentropic expansion (process 3-4)
4. Refrigeration at constant pressure (process 4-1)

For this ideal cycle, assuming an ideal gas with constant specific heat:

$$W_{in} = h_2 - h_1 = Cp(T_2 - T_1)$$

$$W_{out} = h_3 - h_4 = Cp(T_3 - T_4)$$

$$Q_{in} = h_1 - h_4 = Cp(T_1 - T_4)$$

The coefficient of performance is given as:

$$\boxed{\beta_R = \frac{Q_{in}}{|W_{net}|} = \frac{1}{\left(\frac{T_2}{T_1}\right) - 1} = \frac{1}{\left(\frac{T_3}{T_4}\right) - 1} = \frac{1}{r_p^{(K-1)/K} - 1}} \qquad (9.20)$$

where $r_p = P_2/P_1 = P_3/P_4$ is the pressure ratio.

# CHAPTER 10

# MIXTURES AND SOLUTIONS

## 10.1 MOLE FRACTION

$$x_i = \frac{n_i}{n} \qquad (10.1)$$

where  $n = \sum_i n_i$ = Number of moles in the mixture

$n_i$ = Number of moles of component i

## 10.2 MASS FRACTION

$$M_f = \frac{m_i}{m} \qquad (10.2)$$

where  $m_i$ = the mass of component i

$m = \sum_i m_i$ = Mass of mixture

## 10.3 DALTON'S RULE OF PARTIAL PRESSURE

The total pressure of a mixture of ideal gases is equal to the sum of the partial pressures when the partial pressures are determined at the volume and temperature of the mixture, i.e.,

$$P = P_1 + P_2 + P_3 + \ldots + P_n = \sum_i p_i \tag{10.3}$$

## 10.4 AMAGAT-LEDUC RULE OF PARTIAL VOLUME

The total volume of a mixture of ideal gases is equal to the sum of the partial volumes of the constituent gases when the partial volumes are determined at the pressure and temperature of the mixture.

$$V = \sum_i V_i \tag{10.4}$$

$V_i$ is the partial volume

V is the volume of the mixture

## 10.5 EXPRESSIONS FOR PERFECT GASES

The following expressions are valid for a mixture of perfect gases:

1. Entropy $\quad\quad S = \sum_i s_i = \sum_i n_i s_i \tag{10.5}$

2. Internal energy $\quad U = \sum_i m_i u_i, \quad u = \dfrac{\sum m_i u_i}{m} \tag{10.6}$

3. Enthalpy $\quad\quad H = \sum_i m_i h_i, \quad h = \dfrac{\sum m_i h_i}{m} \tag{10.7}$

4. Specific heats $\quad C_v = \dfrac{\sum m_i C_{vi}}{m}, \quad C_p = \dfrac{\sum m_i C_{pi}}{m} \tag{10.8}$

where

$\quad m_i$ = Mass of component i

$\quad m$ = Mass of the mixture

$u_i$ = Internal energy of component i

$u$ = Internal energy of the mixture

$h_i$ = Enthalpy of component i

$h$ = Enthalpy of the mixture

$C_{pi}$ = Constant-pressure specific heat of component i

$C_{vi}$ = Constant-volume specific heat of component i

$P_i$ = Pressure of component i

$T_i$ = Temperature of component i

## 10.6 MIXTURES INVOLVING GASES AND VAPOR

A gas-vapor mixture is an important type of gas mixture from which one or more of the constituent gases can be condensed out. The component that may condense out is called the vapor of the mixture.

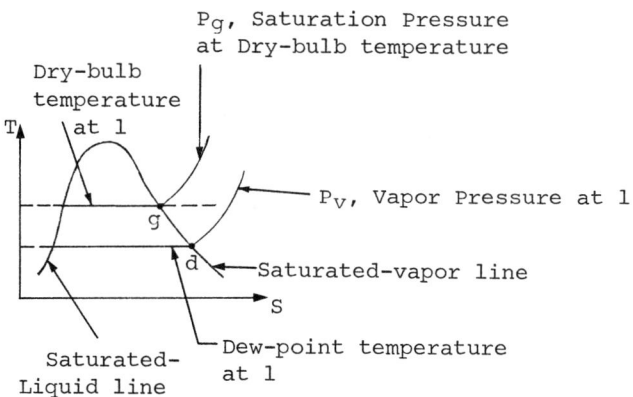

Fig. 10.1 Vapor Pressure, Saturation Pressure, Dew-Point Temperature, and Dry-Bulb Temperature

## 10.6.1 DEW POINT TEMPERATURE

The dew point of a gas vapor mixture is the temperature at which the vapor condenses or solidifies when it is cooled at constant pressure.

a) Saturated air is a mixture of dry air and saturated water vapor.
b) Unsaturated air is a mixture of dry air and superheated vapor.
c) Dry-bulb temperature is the equilibrium temperature of the mixture indicated by an ordinary thermometer.
d) Wet-bulb temperature is the temperature indicated by a wet-bulb thermometer which has been covered with a water-saturated cotton wick.
e) Specific humidity is defined as the ratio of the mass of water vapor to the mass of dry air in a given volume of mixture or

$$\omega = \frac{m_v}{m_a} \quad (10.9)$$

Also:

$$\omega = 0.622 \frac{p_v}{p_a} = 0.622 \frac{p_v}{p - p_v} \quad (10.10)$$

where
$p_v$ = Partial pressure of vapor

$p_a$ = Partial pressure of dry air in the same volume of mixture

$p$ = Pressure of mixture

f) Relative humidity is defined as the ratio of the partial pressure of water vapor in a mixture to the saturation pressure of water at the dry-bulb temperature, or

$$\phi = \frac{p_v}{p_g} \quad (10.11)$$

Also:

$$\phi = \frac{\rho_v}{\rho_g} = \frac{p_a \omega}{0.622 \, p_g} \quad (10.12)$$

where
$\rho_v$ = Partial density of the water vapor

$\rho_g$ = Density of water at the temperature of water

# 10.7 ENTHALPY AND ENTROPY OF A GAS-VAPOR MIXTURE

a) Enthalpy

$$h = h_a + \omega h_v \qquad (10.13)$$

b) Entropy

$$s = s_a + \omega s_v \qquad (10.14)$$

where

$h_a, s_a$ = Specific enthalpy and entropy of the dry air in Btu/lbm of dry air

$h_v, s_v$ = Specific enthalpy and entropy of water vapor

$\omega$ = Specific humidity

$h, s$ = Specific enthalpy and entropy of the mixture

# 10.8 ADIABATIC SATURATION PROCESS

a) The adiabatic saturation process is an important process involving an air-water vapor mixture, in which an air-vapor mixture comes in contact with a body of water in a well-insulated duct (Fig. 10.2):

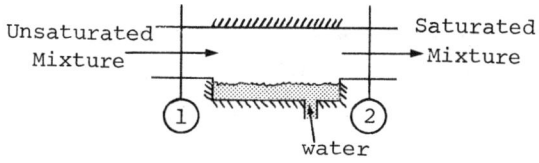

Fig. 10.2  The adiabatic saturation process

The relative humidity and the humidity ratio of the entering air-vapor mixture can be determined from the measurements of the pressure and temperature of the air-vapor mixture entering and leaving the adiabatic saturator. The adiabatic saturation process is one means of determining the humidity of an air-vapor mixture.

b) The humidity of an air-water vapor mixture is usually found from dry-bulb and wet-bulb data. These data are obtained by use of a phychrometer (Fig. 10.3):

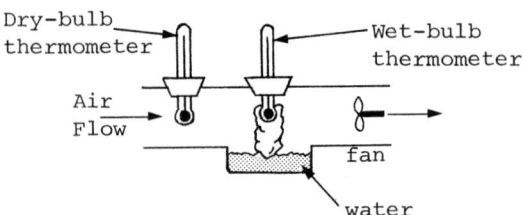

Fig. 10.3 Steady-flow phychrometer for measuring dry and wet temperatures

c) Properties of air-water vapor mixtures are given in graphical form on phychrometric charts. The basic phychrometric chart consists of a plot of dry-bulb temperature (abscissa) and ratio (ordinate).

## 10.9 FUGACITY

The fugacity $f_A$ of component A in a mixture is defined,

$$(d\bar{G}_A)_T = RTd(\ln \bar{f}_A)_T \qquad (10.15)$$

with the requirement that

$$\lim_{p \to 0} \left( \frac{\bar{f}_A}{Y_A P} \right) = 1$$

# 10.10 ACTIVITY

a) The activity $a_A$ of a component A in a mixture at T,P is defined:

$$a_A = \frac{\bar{f}_A}{f_A^0} \qquad (10.16)$$

where $f_A^0$ = Fugacity of pure substance A at $T, P^0$

$P^0$ = Standard state pressure (for gaseous mixtures P = 1 atm)

## 10.10.1 ACTIVITY COEFFICIENT

Another parameter commonly used in the description of mixtures is the activity coefficient $\gamma$, which for any component A, is defined in terms of its activity and mole fraction as

$$\gamma_A = \frac{a_A}{Y_A} \qquad (10.17)$$

# CHAPTER 11

# CHEMICAL REACTIONS

## 11.1 THE COMBUSTION PROCESS

Combustion is a process involving the reaction of a fuel and an oxidizer, in which the stored chemical energy in the fuel is released. A complete combustion reaction can be represented as:

$$C_xH_y + \underbrace{aO_2 + 3.76\, bN_2}_{\text{air}} \rightarrow cCO_2 + dH_2O + eO_2 + fN_2 \quad (11.1)$$

where x,y determine the type of fuel a,b,c,d,e,f are moles of each particular component.

a) Theoretical Air (TA)

The minimum amount of oxygen for the complete combustion of all the elements in the fuel.

b) Excess Air (EA)

The amount of air supplied over and above the theoretical air.

c) Air-Fuel Ratio (AF)

The ratio of the mass of theoretical air to the mass of the fuel.

d) The combustion efficiency $\eta_{comb}$ is defined,

$$\boxed{\eta_{comb} = \frac{F.A.\ \text{Ideal}}{F.A.\ \text{actual}}} \quad (11.2)$$

where F.A. is the fuel-air ratio.

## 11.2 ENTHALPY OF REACTION

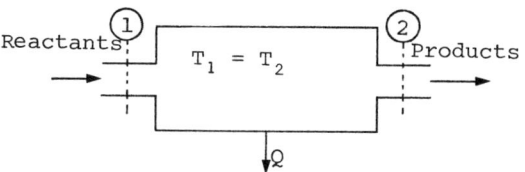

Fig. 11.1 Combustion Process

The concept of the enthalpy of reaction can be understood with the aid of Figure 11.1, where a chemical reaction takes place in a steady-flow process at constant pressure with no work transfer. Neglecting KE and PE and applying the first law of thermodynamics, we have

$$Q = H_2^P - H_1^R = \Delta H \qquad (11.3)$$

$Q$ = Heat flow in
$H_1^R$ = Enthalpy of the reactants at state 1
$H_2^P$ = Enthalpy of the products at state 2

If the reactants and products are both at the same temperature, the quantity DH is called the enthalpy of reaction.

## 11.3 ENTHALPY OF FORMATION ($h_f^o$)

By definition, the standard enthalpy of formation of a compound is the enthalpy of reaction for the formation of the compound from its elements at stable state ($25\,^0C$ and 1 atm).

Notes
a) Exothermic reactions ($\Delta H<0$) are those which liberate heat.
   Endothermic reactions ($\Delta H>0$) are those which absorb heat.
b) The heating value of a fuel is numerically equal to its enthalpy of reaction but with opposite sign:

$$H_1^R - H_2^P = -\Delta H \qquad (11.4)$$

c) $HHV = LHV + m_{H_2O} h_{fg}$ \qquad (11.5)

   HHV = Higher heating value of fuel

LHV = Lower heating value of fuel

$m_{H_2O}$ = Amount of $H_2O$ formed

$h_{fg}$ = Enthalpy of vaporization of $H_2O$

d) The enthalpy of formation of an element is zero.

e) The total molal enthalpy at any temperature and pressure $\bar{h}_{T,P}$ is:

$$\bar{h}_{T,P} = \bar{h}_f^0 + (\bar{h}_{T,P} - \bar{h}_{298(atm)}) \quad (11.6)$$

where $\bar{h}_{T,P} - h_{298\cdot(atm)}$ = Difference in enthalpy between any given state and the enthalpy at the reference state of $298\,^0K$ and 1 atm

$\bar{h}_f^0$ = Enthalpy of formation of a substance

f) Standard enthalpy of reaction, $\Delta H^0 = H_2^0 - H_1^0$

g) The Gibbs function of formation, $g_f^0$, has been defined by a procedure similar to the enthalpy of formation.

# 11.4 FIRST LAW ANALYSIS OF REACTING SYSTEMS

Applying the first law to steady-state, steady-flow processes involving a chemical reaction and negligible changes in kinetic and potential energy we can write:

$$Q_{cv} - W_{cv} = \sum_P n_e (\bar{h}_f + \Delta \bar{h})_e - \sum_R n_i (\bar{h}_j + \Delta \bar{h})_i \quad (11.7)$$

where

a) R, P refer to the reactants and products respectively

b) $n_i$ = Number of reactants, $n_e$ = Number of products

c) $\bar{h}_j$ = Molal enthalpy of formation

d) $\Delta \bar{h} = \bar{h}_f^0 - \bar{h}_{298}^0$ of the substance

Notes

1. $\Delta \bar{h}$ can be found directly from tables.

2. If the deviation from ideal gas behavior is significant but no tables are available, the value for $\Delta \bar{h}$ can be found from the generalized charts and the values for $Cp_0$ or $\Delta \bar{h}$ at 0.1 MPa pressure.

## 11.5 ADIABATIC FLAME TEMPERATURE

For any process that takes place adiabatically and with no work or changes in kinetic or potential energy, the temperature of the products is referred to as the adiabatic flame temperature or adiabatic combustion temperature.

## 11.6 THE THIRD LAW OF THERMO-DYNAMICS AND ABSOLUTE ENTROPY

The third law of thermodynamics states that the entropy of a pure crystalline substance is zero at the absolute zero of temperature ($0\,^0K$). The important result of this law is that it permits the determination of the absolute entropy for ideal-gas mixtures which is valid for most practical problems as follows:

$$\bar{S}_{T,P} = \bar{S}_T^0 - \bar{R} \ln p \qquad (11.8)$$

where

$\bar{S}_{T,P}$ = Absolute entropy at 0.1 MPa and temperature T

P = Pressure expressed in atmospheres
$\bar{R}$ = Universal gas constant

## 11.7 SECOND LAW ANALYSIS OF REACTING SYSTEMS

The second law for any reactive process may be written:

$$\boxed{\Delta S = S_P - S_R - \Sigma \frac{Q_{cv}}{T} \geq 0} \qquad (11.9)$$

where  $S_P$ = Entropy of the products

$S_R$ = Entropy of the reactants

$\Sigma \dfrac{Q_{cv}}{T}$ = Entropy transfer into control volume due to heat transfer

The concepts of reversible work, irreversibility and availability are applied to the reacting systems. The basic formulas are listed below (SSSF) (in the absence of kinetic and potential energy, changes):

a) Reversible work  (11.10)

$$W_{rev} = \Sigma_R n_i \left[ h_j^0 + \Delta \bar{h} - T_0 \bar{s} \right]_i - \Sigma_P n_e \left[ h_f^0 + \Delta \bar{h} - T_0 \bar{s} \right]_e$$

b) Irreversibility

$$I = \Sigma_P n_e T_0 \bar{s}_e - \Sigma_R n_i T_0 \bar{s}_i - Q_{cv} \qquad (11.11)$$

c) Availability

$$\psi = (h - T_0 s) - (h_0 - T_0 s_0) \qquad (11.12)$$

## 11.8 GIBB'S FUNCTION

The Gibbs function is defined:

$$\boxed{G = H - TS} \qquad (11.13)$$

where    H = Enthalpy
            S = Entropy
            T = Temperature

For a chemical reaction carried out at constant temperature and pressure we have,

$$\boxed{\Delta G = \Delta H - T \Delta S \leq 0} \qquad (11.14)$$

This means that a chemical reaction is possible only if the Gibbs function for the products is less than the Gibbs function for the reactants.

## 11.9 CHEMICAL POTENTIAL

Consider a homogeneous mixture. If there are i constituents, we may express the internal energy in functional form as

$$U = f(s, v, n_1, n_2, \ldots, n_i).$$

A differential change in the internal energy is given,

$$(11.15)$$

$$dU = \left(\frac{\partial U}{\partial s}\right)_{v, N_i} dS + \left(\frac{\partial U}{\partial V}\right)_{s, N_i} dV + \Sigma \left(\frac{\partial U}{\partial N_i}\right)_{s, v, N} dN_i$$

The partial derivatives in the summation of the equation are defined as the chemical potential $u_i$. Also we have the equivalent relations for the chemical potential:

$$u_i = \left(\frac{\partial H}{\partial N_i}\right)_{s,p,N_j} = \left(\frac{\partial A}{\partial N_i}\right)_{T,V,N} = \left(\frac{\partial G}{\partial N_i}\right)_{T,P,N_j} \quad j \neq i$$

(11.16)

where      H = Enthalpy = u + pv
               A = Helmholtz function = U - TS
               G = Gibbs function = H - TS

# CHAPTER 12

# CHEMICAL EQUILIBRIUM

## 12.1 REQUIREMENTS FOR CHEMICAL EQUILIBRIUM

Applying the Gibbs function to a reactive system, we find that a chemical reaction carried out at constant pressure and temperature can proceed only if the Gibbs function of the system will continually decrease. The reaction will stop where the Gibbs function of the system has reached a minimum (Fig. 12.1).

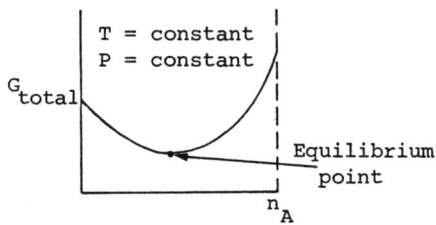

Fig. 12.1 Illustration of the requirement for the chemical equilibrium

We thus see that the equilibrium composition of any reactive system of known temperature and pressure is governed by

$$dG_{T,P} = 0 \qquad (12.1)$$

## 12.2 EQUILIBRIUM AND CHEMICAL POTENTIAL

To find the equilibrium composition of a mixture of gases undergoing a chemical reaction, we need an expression for dG in terms of the moles of reactants and products present at any given time:

$$dG = VdP - SdT + \sum_i u_i \, dN_i \qquad (12.2)$$

where

$$u_i = \left( \frac{\partial G}{\partial N_i} \right)_{P,T,N_i}$$

$N_i$: Number of moles of each chemical species within the system at some time.

## 12.3 EQUILIBRIUM CONSTANT OF A REACTIVE MIXTURE OF IDEAL GASES

We define the equilibrium constant $k_p$ for ideal gas reactions by the relation:

$$K = \frac{(P_C)^{V_C} (P_D)^{V_D}}{(P_A)^{V_A} (P_B)^{V_B}} \qquad (12.3)$$

where

$P_A$, $P_B$, $P_C$, $P_D$ are the partial pressures of the chemical constitutents

$V_A$, $V_B$, $V_C$, $V_D$ are the stoichiometric coefficients.

Other forms of the equilibrium constant are:

a) 
$$\ln K = - \frac{\Delta G^0}{RT} \qquad (12.4)$$

where

$$\Delta G^0 = V_C \bar{g}_C^0 + V_D \bar{g}_D^0 - V_A \bar{g}_A^0 - V_B \bar{g}_B^0$$

$\bar{g}_C^0$, $\bar{g}_D^0$, $\bar{g}_A^0$, $\bar{g}_B^0$ are the standard Gibbs functions.

T = Temperature of the environment
$\overline{R}$ = Universal gas constant

b)
$$K = \frac{a_C^{V_C} \cdot a_D^{V_D}}{a_A^{V_A} \cdot a_B^{V_B}} \qquad (12.5)$$

$a_A$, $a_B$, $a_C$, $a_D$ are the activity coefficients.

c)
$$K = \frac{x_C^{V_C} \, x_D^{V_D}}{x_A^{V_A} \, x_B^{V_B}} \, (P)^{V_C+V_D-V_A-V_B} \qquad (12.6)$$

$x_A$, $x_B$, $x_C$, $x_D$ are the mole fractions.

(P) is the mixture pressure.

## 12.4 EQUILIBRIUM BETWEEN TWO PHASES OF A PURE SUBSTANCE

Under equilibrium conditions the Gibbs function of each phase of a pure substance is equal.

## 12.5 EQUILIBRIUM OF A MULTICOM-PONENT, MULTIPHASE SYSTEM

The requirement for equilibrium is that the chemical potential of each component is the same in all phases.

## 12.6 GIBB'S PHASE RULE

a) For an open system we have:

$$\boxed{F = n + 2 - r} \quad (12.7)$$

where  n = Number of components
r = Number of phases
F = Number of independent intensive properties

b) For a closed system we have:

$$\boxed{\begin{array}{c} F = n + 2 - r \\ n \leq r \end{array}} \quad (12.8)$$

# CHAPTER 13

# FLOW THROUGH NOZZLES AND BLADE PASSAGES

## 13.1 CONSERVATION OF MASS FOR THE CONTROL VOLUME

$$\frac{\partial}{\partial t} \int_V p\,dV + \int_A p\vec{V} \cdot d\vec{A} = 0 \qquad (13.1)$$

where

$\frac{\partial}{\partial t} \int_V p\,dV$ = Rate of change of mass within the control volume

$\int_A p\bar{V}dA$ = Net rate of mass efflux through the control surface

### 13.1.1 SPECIAL CASES

For an incompressible flow (p=constant), equation (13.1) becomes

$$\int_A p\bar{V} \cdot d\bar{A} = 0 \qquad (13.2)$$

## 13.2 MOMENTUM EQUATION FOR THE CONTROL VOLUME

$$\Sigma F_j = \frac{1}{g_c}\left[\frac{d}{dt}\int_V V_j \rho\,dV + \int_A V_j \rho V_{rn}\,dA\right] \qquad (13.3)$$

For the steady-state, steady-flow process we have:

$$\Sigma F_j = \frac{1}{g_c}\left[\Sigma \dot{m}_e (v_e)_j - \Sigma \dot{m}_i (v_i)_j\right] \quad (13.4)$$

where

$\dot{m}_i, \dot{m}_e$ = Rate of mass entering and leaving the C.V.

$v_i, v_e$ = Velocity of the mass entering and leaving the C.V.

$i$ = x,y,z (directions)

## 13.3 SPEED OF SOUND

### 13.3.1 MACH NUMBER

The Mach number is defined: $M = \frac{V}{C}$ \quad (13.5)

where  V = Local flow speed
c = Local speed of sound

For an ideal gas $c = \sqrt{kRT}$, where $k = \frac{C_p}{C_v}$ :

if M > 1, the flow is supersonic.
if M = 1, the flow is sonic.
if M < 1, the flow is subsonic.

## 13.4 LOCAL ISENTROPIC STAGNATION PROPERTIES

Local isentropic stagnation properties are those properties that would be obtained at any point in a flow field if the fluid at that point were decelerated from local conditions to zero velocity following a frictionless adiabatic - that is, isentropic - process.

For an ideal gas the isentropic stagnations (denoted by the subscript o) are:

$$\frac{P_0}{P} = \left[1 + \frac{k-1}{2} M^2\right]^{k/(k-1)} \qquad (13.6)$$

$$\frac{T_0}{T} = 1 + \frac{k-1}{2} M^2 \qquad (13.7)$$

$$\frac{\rho_0}{\rho} = \left[1 + \frac{k-1}{2} M^2\right]^{1/(k-1)} \qquad (13.8)$$

## 13.5 CRITICAL CONSTANTS

The conditions at the throat of the nozzle can be found by noting that $M = 1$ at the throat. These properties (denoted by an asterisk *) are referred to as critical pressure, critical temperature and critical density and the ratios are given below:

$$\frac{P^*}{P_0} = \left(\frac{2}{k+1}\right)^{k/(k-1)} \qquad (13.9)$$

$$\frac{T^*}{T_0} = \left(\frac{2}{k+1}\right) \qquad (13.10)$$

$$\frac{\rho^*}{\rho_0} = \left(\frac{2}{k+1}\right)^{1/(k-1)} \qquad (13.11)$$

## 13.6 EFFECTS OF AREA VARIATION ON FLOW PROPERTIES IN ISENTROPIC FLOW

In considering the effect of area variation on flow properties in isentropic flow, the following equation may be used:

$$\boxed{\frac{dA}{A} = \frac{dP}{\rho V^2}(1 - M^2)}\qquad (13.12)$$

where   $M$ = Mach number

$dA$ = Change in the area

From this equation we can draw the following conclusions (Fig. 13.1) about the proper shape for nozzles and diffusers:

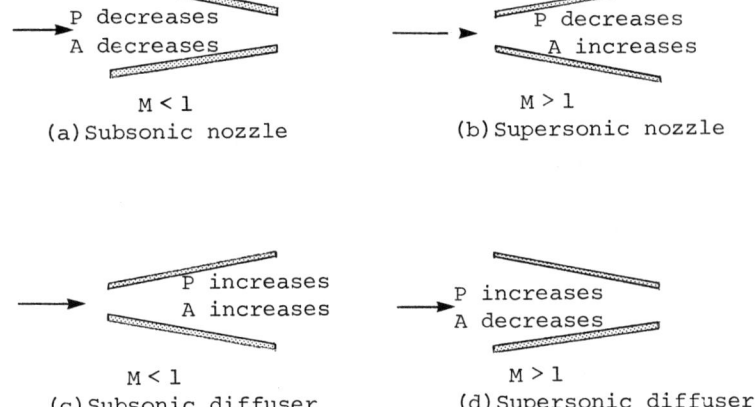

Fig. 13.1 Required area changes for nozzles and diffusers

## 13.7 ISENTROPIC FLOW OF AN IDEAL GAS

For the isentropic flow of an ideal gas we can summarize the basic equations as follows:

Continuity:   $\rho_1 V_1 A_1 = \rho_2 V_2 A_2 = \rho V A = \dot{m}$   (13.13)

Momentum:   $R_x + p_1 A_1 - p_2 A_2 = \dot{m} V_2 - \dot{m} V_1$   (13.14)

First law:   $h_1 + \dfrac{V_1^2}{2} = h_2 + \dfrac{V_2^2}{2} = h + \dfrac{V^2}{2}$   (13.15)

Second law: $s_1 = s_2 = s$ (13.16)

Equation of state: $p = \rho RT$ (13.17)

Process equation: $p/\rho^k = \text{constant}$ (13.18)

## 13.8 ISENTROPIC FLOW IN A CONVERGING AND A CONVERGING / DIVERGING NOZZLE

### 13.8.1 CONVERGING NOZZLE

Flow through the converging nozzle shown in Fig. 13.2 is supplied from a large chamber, where the conditions are assumed to be stagnation conditions. The flow is induced by a vacuum pump downstream and is controlled by the valve shown. The back pressure $p_b$ to which the nozzle discharges is controlled by the valve. The upstream stagnation conditions ($p_0, T_0$, etc.) are maintained constant. The pressure in the exit plane of the nozzle is denoted $p_e$.

The effect of variations in $p_b$ on the pressure distribution through the nozzle, on the mass flow rate, and on the exit plane pressure are illustrated graphically in Fig. 13.2:

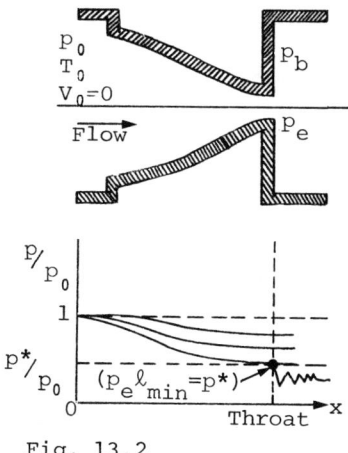

Fig. 13.2

## 13.8.2 CONVERGING-DIVERGING NOZZLE

As in the previous case, the flow through the converging-diverging nozzle is induced by a vacuum pump, and is controlled by the valve shown. The upstream stagnation conditions are assumed constant; the pressure in the exit plane of the nozzle is denoted $p_e$; the nozzle discharges to the back pressure $p_b$. The effects of variations in back pressure on the pressure distribution through the nozzle are illustrated graphically in Fig. 13.3:

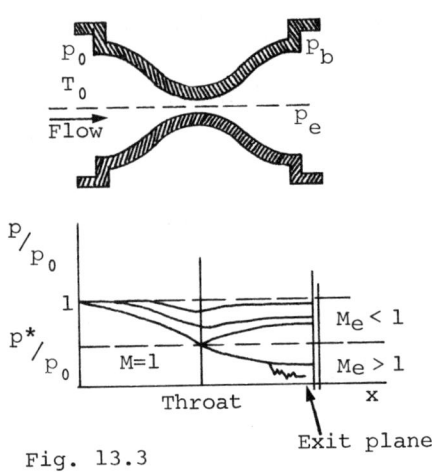

Fig. 13.3

## 13.9 NORMAL SHOCKS

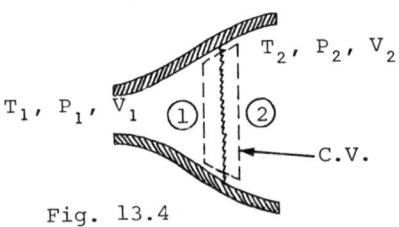

Fig. 13.4

a) A shock wave involves an extremely rapid and abrupt change of state. Figure 13.4 shows a control volume that can be used to analyze such a normal shock. The basic equations applied to the thin control volume shown in Fig. 13.4 are:

Continuity: $\quad \rho_1 V_1 = \rho_2 V_2 = G \quad (13.19)$

Momentum Equation: $\quad P_1 + \rho_1 V_1^2 = P_2 + \rho_2 V_2^2 \quad (13.20)$

First Law: $\quad h_1 + \dfrac{V_1^2}{2} = h_2 + \dfrac{V_2^2}{2} \quad (13.21)$

Enthalpy: $\quad h_{01} = h_{02} \quad (13.22)$

Second Law: $\quad s_2 - s_1 = C_p \ln \dfrac{T_2}{T_1} - R \ln \dfrac{P_2}{P_1} \quad (13.23)$

Equation of State: $\quad P = \rho RT \quad (13.24)$

Property changes across a normal shock are summarized in table 13.1:

| Property | Effect | Obtained from |
|---|---|---|
| Stagnation temperature, $T_0$ | Constant | Energy equation |
| Entropy, s | Increase | Second Law |
| Stagnation pressure, $p_0$ | Decrease | Ts diagram |
| Temperature, T | Increase | Ts diagram |
| Velocity, V | Decrease | Energy equation, and effect on T |
| Density, $\rho$ | Increase | Continuity equation, and effect on V |
| Pressure, p | Increase | Momentum equation, and effect on V |
| Mach number, M | Decrease | $M = V/c$, and effect on V and T |

Table 13.1

## 13.9.1 FLOW IN A CONVERGING-DIVERGING NOZZLE

Since we have considered normal shocks, we are now in a position to complete our discussion of flow in a converging-diverging nozzle. The pressure distribution through the nozzle for different back pressures is shown in Fig. 13.5.

In Regime 1 the flow is subsonic throughout. At condition (iii), the flow at the throat is sonic, that is, $M_t = S$.

In Regime 2 the exit flow is subsonic, and consequently $p_e = p_b$.

In Regime 3 the back pressure is higher than the exit pressure but not sufficiently high to sustain a normal shock in the exit plane.

In Regime 4 the flow adjusts to the lower back pressure through a series of oblique expansion waves.

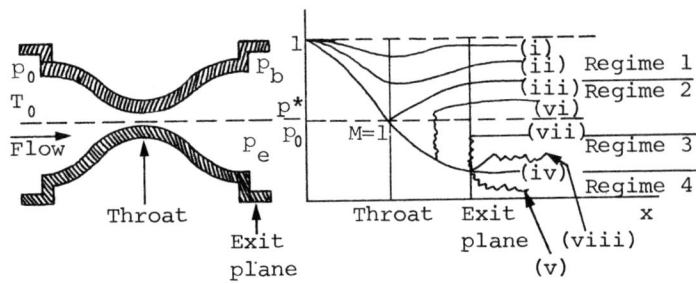

Fig. 13.5

## 13.10 NOZZLE AND DIFFUSER COEFFICIENTS

a) Nozzle efficiency is defined:

$$n_N = \frac{\text{Actual kinetic energy at nozzle exit}}{\text{Kinetic energy at nozzle exit with isentropic flow to same exit pressure}} \quad (13.25)$$

b) The coefficient of discharge $C_p$ is defined by the relation:

$$C_p = \frac{\text{Actual mass rate of flow}}{\text{Mass rate of flow with isentropic flow}} \qquad (13.26)$$

c) The efficiency of a diffuser is defined:

$$n_p = \frac{(1 + \frac{k-1}{2} M_1^2)\left[\left(\frac{P_{02}}{P_{01}}\right)^{(k-1)/K} - 1\right]}{\frac{k-1}{2} M_1^2} \qquad (13.27)$$

where

1. states 1 and 01 are the actual and stagnation states of the fluid entering the diffuser
2. states 2 and 02 are the actual and stagnation states of the fluid leaving the diffuser

d) The velocity coefficient $C_v$ is defined:

$$C_v = \frac{\text{Actual velocity at nozzle exit}}{\text{Velocity at nozzle exit with isentropic flow and same exit pressure}} \qquad (13.28)$$

## 13.11 FLOW THROUGH BLADE PASSAGES

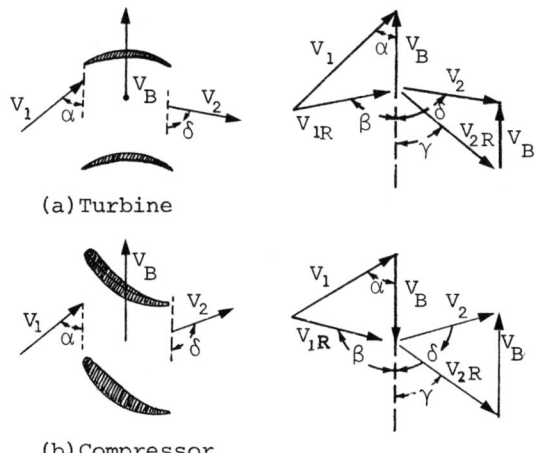

(a) Turbine

(b) Compressor

Fig. 13.6 Velocity vector diagrams

Figure 13.6 shows vector diagrams for both a turbine and a compressor.

In the diagrams:

a) $V_1$, $V_{1R}$ represents the velocity and relative velocity of the fluid entering at angles $\alpha$ and $\beta$, respectively.

b) $V_2$, $V_{2R}$ represents the velocity and relative velocity of the fluid leaving at angles $\delta$ and $\gamma$, respectively.

Applying the first law, assuming steady adiabatic flow, for a stationary and moving observer we have:

$$W = \frac{(V_1^2 - V_{1R}^2) - (V_2^2 - V_{2R}^2)}{2} \qquad (13.29)$$

where W is the work done on the blade.

Applying the second law to this process we conclude that

$$s_2 \geq s_1 \qquad (13.30)$$

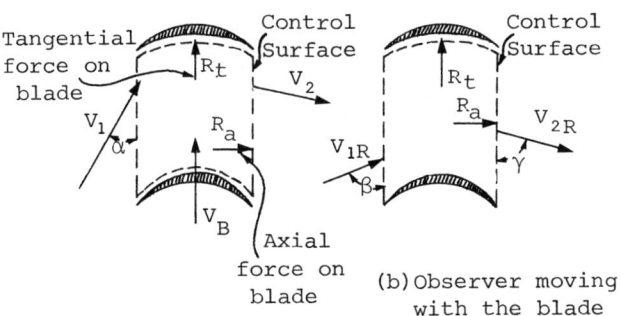

Fig. 13.7  Analysis of forces on a turbine

If we apply the momentum equation in the tangential direction for both cases (see Fig. 13.6) we have:

a) For a stationary observer,

$$R_t = \dot{m}(V_1 \cos \alpha + V_2 \cos \delta) \qquad (13.31)$$

$$R_a = \dot{m}(V_2 \sin\delta - V_1 \sin\alpha) - (P_1A_1 - P_2A_2) \quad (13.32)$$

b) For a moving observer,

$$R_t = \dot{m}(V_{1R}\cos\beta + V_{2R}\cos\gamma) \quad (13.33)$$

$$R_a = \dot{m}(V_{2R}\sin\gamma - V_{1R}\sin\beta) - (P_1A_1 - P_2A_2) \quad (13.34)$$

where $\dot{m}$ represents the mass flow rate out and in of the control volume.

## 13.12 IMPULSE AND REACTION STAGES FOR TURBINES

### 13.12.1 IMPULSE STAGE

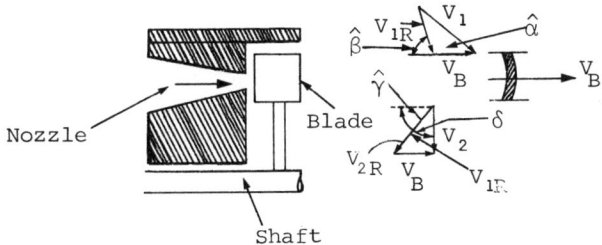

Fig. 13.8 Impulse stage

In an impulse stage, the entire pressure drop taken place in a stationary nozzle, and the pressure remains constant as the fluid flows through the blade passage. There is a decrease in the kinetic energy of the fluid as it flows through the blade passage, and the enthalpy will increase because of reversibilities associated with the fluid flow.

The blade efficiency $\eta_B$ is defined:

$$\eta_B = \frac{W}{V_1^2/2} = 2\frac{V_B}{V_1}\left(\cos\alpha - \frac{V_B}{V_1}\right)\left[\frac{1 + \dfrac{K_B \cos\gamma}{\cos\alpha - V_B/V_1}}{\sqrt{(\cos - V_B/V_1)^2 + \sin^2\alpha}}\right]$$

$$(13.35)$$

where

$$K_B = \frac{V_{2R}}{V_{1R}}, \text{ blade velocity coefficient}$$

$$\eta_B = f(V_B/V_1, \alpha, \gamma, K_B)$$

## 13.12.2 REACTION STAGE

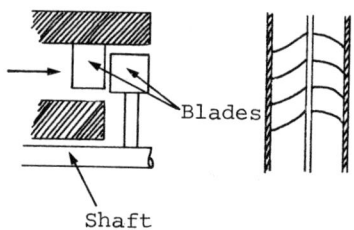

Fig. 13.9   Reaction stage

In the pure reaction stage, the entire pressure drop occurs as the fluid flows through the moving blades. Thus the moving blade acts as a nozzle, and the blade passage must have the proper contour for a nozzle. In the pure reaction stage the only purpose of the stationary blade is to direct the fluid into the moving blade at the proper angle and velocity.

A comparison of the impulse stage with the reaction stage regarding the blade speed ratio for maximum efficiency yields:

$$\frac{V_B}{V_1} = \frac{V_B}{V_0} = 0.5 \text{ for the impulse stage} \qquad (13.36)$$

$$\frac{V_B}{V_1} = \frac{V_B}{V_0} = \frac{1}{\sqrt{2}} \text{ for the reaction stage} \qquad (13.37)$$

where   $V_0 = \sqrt{2g_c Dh_s}$ \qquad (13.38)

### THE BEST AND MOST COMPREHENSIVE IN TEST PREPARATION

# COLLEGE BOARD ACHIEVEMENT TEST

# CHEMISTRY

- Based on the most recent exams.
- Six full length exams.
- Each exam is 1 hour.
- Solution methods to questions are extensively illustrated in detail.
- Almost every type of question that can be expected on the Chemistry Achievement Test.
- Complete Answer Key follows every exam.
- Enables students to discover their strengths and weaknesses and thereby become better prepared.

Available at your local bookstore or order directly from us by sending in coupon below.

---

**RESEARCH and EDUCATION ASSOCIATION**
505 Eighth Avenue   New York, N.Y. 10018
Phone: (212) 695-9487

Please check one box:
☐ Check enclosed
☐ Visa
☐ MasterCard

Charge Card Number ☐☐☐☐☐☐☐☐☐☐☐☐☐

Expiration Date (Mo./Yr.) _____

Please ship the "CBAT Chemistry" @ $12.95 plus $2.00 for shipping.

Name.....................................................
Address..................................................
City............................................State.................

THE BEST AND MOST COMPREHENSIVE IN TEST PREPARATION

# MCAT

## MEDICAL COLLEGE ADMISSION TEST

- Based on the most recent exams.
- Six full length exams. Over 750 pages.
- Each exam is $6\frac{1}{2}$ hours.
- Solution methods to questions are extensively explained in detail.
- Almost every type of question that can be expected on an MCAT.
- Complete Answer Key follows every exam.
- Enables students to discover their strengths and weaknesses and thereby become better prepared.

Available at your local bookstore or order directly from us by sending in coupon below.

---

**REA** RESEARCH and EDUCATION ASSOCIATION
505 Eighth Avenue  New York, N.Y. 10018
Phone: (212) 695-9487

VISA    MasterCard

Please check one box:
☐ Check enclosed
☐ Visa
☐ MasterCard

Charge Card Number ☐☐☐☐☐☐☐☐☐☐☐☐☐☐

Expiration Date (Mo./Yr.) _____

Please ship the MCAT @ $17.85 plus $2.00 for shipping.

Name..................................................
Address...............................................
City.........................................State............

# THE ENGLISH HANDBOOK
## OF
# GRAMMAR, STYLE,
## AND
# COMPOSITION

- This book illustrates the rules and numerous exceptions that are characteristic of the English language, in great depth, detail, and clarity.

- Over 2,000 examples comparing correct and wrong usage in all areas of grammar and writing.

- Solves the usual confusion about punctuation.

- Illustrates spelling "tricks" and how to remember correct spelling.

- Teaches how to acquire good writing skills.

- Provides special learning exercises at the end of each chapter to prepare for homework and exams.

- Fully indexed for locating specific topics rapidly.

Available at your local bookstore or order directly from us by sending in coupon below.

---

**RESEARCH and EDUCATION ASSOCIATION**
505 Eighth Avenue   New York, N.Y. 10018
Phone: (212) 695-9487

Please check one box:
- ☐ Check enclosed
- ☐ Visa
- ☐ MasterCard

Charge Card Number ☐☐☐☐☐☐☐☐☐☐☐☐
Expiration Date (Mo./Yr.)_____

Please ship the "English Handbook" @ $12.95 plus $2.00 for shipping.

Name.....................................................................
Address..................................................................
City.............................................State....................

# HANDBOOK AND GUIDE FOR COMPARING and SELECTING COMPUTER LANGUAGES

| | |
|---|---|
| **BASIC** | **PL/1** |
| **FORTRAN** | **APL** |
| **PASCAL** | **ALGOL-60** |
| **COBOL** | **C** |

- This book is the first of its kind ever produced in computer science.

- It examines and highlights the differences and similarities among the eight most widely used computer languages.

- A practical guide for selecting the most appropriate programming language for any given task.

- Sample programs in all eight languages are written and compared side-by-side. Their merits are analyzed and evaluated.

- Comprehensive glossary of computer terms.

Available at your local bookstore or order directly from us by sending in coupon below.

---

**RESEARCH and EDUCATION ASSOCIATION**
505 Eighth Avenue  New York, N.Y. 10018
Phone: (212)695-9487

Please check one box:
☐ Check enclosed
☐ Visa
☐ MasterCard

Charge Card Number ☐☐☐☐☐☐☐☐☐☐☐☐☐☐

Expiration Date (Mo./Yr.) _____

Please ship the "Computer Languages Handbook" @ $8.95 plus $2.00 for shipping.

Name................................................
Address.............................................
City................................State............

# HANDBOOK of MATHEMATICAL, SCIENTIFIC, and ENGINEERING

## FORMULAS, TABLES, FUNCTIONS, GRAPHS, TRANSFORMS

A particularly useful reference for those in math, science, engineering and other technical fields. Includes the most-often used formulas, tables, transforms, functions, and graphs which are needed as tools in solving problems. The entire field of special functions is also covered. A large amount of scientific data which is often of interest to scientists and engineers has been included.

Available at your local bookstore or order directly from us by sending in coupon below.

---

**RESEARCH and EDUCATION ASSOCIATION**
505 Eighth Avenue  New York, N.Y. 10018
Phone: (212)695-9487

VISA   MasterCard

Please check one box:
☐ Check enclosed
☐ Visa
☐ MasterCard

Charge Card Number ☐☐☐☐☐☐☐☐☐☐☐☐☐☐

Expiration Date (Mo./Yr.)_____

Please ship the "Math Handbook" @ $21.85 plus $2.00 for shipping.

Name.................................................................
Address..............................................................
City............................................State................

# THE PROBLEM SOLVERS

The "PROBLEM SOLVERS" are comprehensive supplemental textbooks designed to save time in finding solutions to problems. Each "PROBLEM SOLVER" is the first of its kind ever produced in its field. It is the product of a massive effort to illustrate almost any imaginable problem in exceptional depth, detail, and clarity. Each problem is worked out in detail with step-by-step solution, and the problems are arranged in order of complexity from elementary to advanced. Each book is fully indexed for locating problems rapidly.

ADVANCED CALCULUS
ALGEBRA & TRIGONOMETRY
AUTOMATIC CONTROL SYSTEMS/ROBOTICS
BIOLOGY
BUSINESS, ACCOUNTING, & FINANCE
CALCULUS
CHEMISTRY
COMPLEX VARIABLES
COMPUTER SCIENCE
DIFFERENTIAL EQUATIONS
ECONOMICS
ELECTRICAL MACHINES
ELECTRIC CIRCUITS
ELECTROMAGNETICS
ELECTRONIC COMMUNICATIONS
ELECTRONICS
FINITE and DISCRETE MATH
FLUID MECHANICS/DYNAMICS
GENETICS

GEOMETRY:
PLANE • SOLID • ANALYTIC
HEAT TRANSFER
LINEAR ALGEBRA
MACHINE DESIGN
MECHANICS: STATICS • DYNAMICS
NUMERICAL ANALYSIS
OPERATIONS RESEARCH
OPTICS
ORGANIC CHEMISTRY
PHYSICAL CHEMISTRY
PHYSICS
PRE-CALCULUS
PSYCHOLOGY
STATISTICS
STRENGTH OF MATERIALS & MECHANICS OF SOLIDS
TECHNICAL DESIGN GRAPHICS
THERMODYNAMICS
TRANSPORT PHENOMENA:
MOMENTUM • ENERGY • MASS
VECTOR ANALYSIS

If you would like more information about any of these books, complete the coupon below and return it to us or go to your local bookstore.

---

**REA** RESEARCH and EDUCATION ASSOCIATION
505 Eighth Avenue • New York, N.Y. 10018
Phone: (212) 695-9487

Please send me more information about your Problem Solver Books.

Name ............................................................
Address .........................................................
City ........................................ State ..............